THE EXPERIMENT

BECAUSE EVERY BOOK IS A TEST OF NEW IDEAS

AN ILLUSTRATED BOOK OF

BAD ARGUMENTS

ALI ALMOSSAWI

ILLUSTRATIONS BY

Alejandro Giraldo

THE EXPERIMENT
NEW YORK

AN ILLUSTRATED BOOK OF BAD ARGUMENTS (second edition)
Creative Commons Ⓒ 2013 Ali Almossawi
All material new to this edition copyright © 2014 Ali Almossawi

This is a revised and updated second edition of *An Illustrated Book of Bad Arguments* (originally published by the author in 2013 at BookofBadArguments.com, under a Creative Commons Attribution-NonCommercial 3.0 Unported license).

The Experiment, LLC
220 East 23rd Street, Suite 301
New York, NY 10010-4674
www.theexperimentpublishing.com

The Experiment's books are available at special discounts when purchased in bulk for premiums and sales promotions as well as for fund-raising or educational use. For details, contact us at info@theexperimentpublishing.com.

Library of Congress Cataloging-in-Publication Data:

Almossawi, Ali, author.
An illustrated book of bad arguments / Ali Almossawi ; illustrations by Alejandro Giraldo.
    pages cm
Includes bibliographical references.
ISBN 978-1-61519-225-0 (cloth) -- ISBN 978-1-61519-226-7 (ebook)
1. Reasoning. 2. Fallacies (Logic) 3. Logic. I. Title.
BC177.A46 2014
168--dc23

                    2014016343

ISBN 978-1-61519-225-0
Ebook ISBN 978-1-61519-226-7

Cover design, art direction, and original text design by Ali Almossawi
Illustrations by Alejandro Giraldo
Additional text design by Pauline Neuwirth, Neuwirth & Associates, Inc.
Back cover design by Karen Giangreco

Manufactured in China
Distributed by Workman Publishing Company, Inc.
Distributed simultaneously in Canada by Thomas Allen & Son Ltd.

First printing August 2014
10 9 8 7 6

To Danah—everything

CONTENTS

The first principle is that you must not fool yourself
and you are the easiest person to fool.

—*Richard P. Feynman*

WHO IS THIS BOOK FOR?

This book is aimed at newcomers to the field of logical reasoning, particularly those who, to borrow a phrase from Pascal, are so made that they understand best through visuals. I have selected nineteen of the most common errors in reasoning and visualized them using memorable illustrations that are supplemented with lots of examples. My hope is that the reader will learn from these pages some of the most common pitfalls in arguments and be able to identify and avoid them in practice.

The literature on logic and logical fallacies is wide and exhaustive. Some of it aims to help the reader utilize the tools and paradigms that support good reasoning, and hence lead to more constructive debates. But reading about things that one should *not* do is also a useful learning experience. In his book *On Writing*, Stephen King writes, "One learns most clearly what not to do by reading bad prose." He describes his experience of reading a particularly terrible novel as "the literary equivalent of a smallpox vaccination" [King]. The mathematician George Pólya is quoted as having said in a lecture about teaching math that, in addition to understanding it well, one must also know how to misunderstand it [Pólya]. This work primarily talks about things that one should *not* do in arguments.[1]

This book's novelty also lies in its use of lively illustrations to describe some of the common errors in reasoning that plague a lot of our present discourse. They are inspired in part by allegories such as Orwell's *Animal Farm* and by the humorous nonsense of works such as Lewis Carroll's stories and poems. Unlike in such works, there isn't a narrative that ties the illustrations together; they are discrete scenes, connected only by style and theme, which better affords adaptability and reuse. Each

---

[1] For a look at the converse, see T. Edward Damer's book *Attacking Faulty Reasoning*.

of the fallacies has just one page of explanation, which I hope will make them easy to digest and remember.

• • •

Many years ago, I spent part of my time writing software specifications using first-order predicate logic. It was an intriguing way of reasoning, using mathematics rather than the usual notation—English. It brought precision where there was potential ambiguity and rigor where there had formerly been some hand-waving.

During the same time, I perused a few books on propositional logic, both modern and medieval, one of which was Robert Gula's *Nonsense: A Handbook of Logical Fallacies* [Gula]. That book reminded me of a list of guidelines that I had scribbled down in a notebook a decade ago about how to argue; they were the result of several years of arguing with strangers in online forums and they included, for example, "Try not to make general claims about things." That is obvious to me now, but to a schoolboy, it was an exciting realization.

It quickly became evident to me that formalizing one's reasoning could lead to useful benefits such as clarity of thought and expression, improved objectivity, and greater confidence. The ability to analyze others' arguments can also serve as a yardstick for when to withdraw from discussions that will most likely be futile.

Issues and events that affect our lives and the societies we live in, such as civil liberties and presidential elections, often cause people to debate policies and beliefs. Observing some of that discourse, one gets the feeling that a noticeable amount of it suffers from the absence of good reasoning.

Of course, logic is not the only tool used in debate, and it is helpful to be cognizant of the others. Rhetoric likely tops the list, followed by concepts such as the "burden of proof" and Occam's razor (the principle that, when seeking to explain a phenomenon, one should not introduce any more conjecture than is needed, also known as the *principle of parsimony*). The interested reader may wish to refer to the wide literature on the topic.

In closing, the rules of logic are not laws of the natural world, nor do they constitute all of human reasoning. As Marvin Minsky asserts, ordinary, commonsense reasoning is difficult to explain in terms of logical principles, as are analogies. He adds, "Logic no more explains how we think than grammar explains how we speak" [Minsky]. Logic does not generate new truths, but rather allows one to evaluate existing chains of thought for consistency and coherence. It is precisely for that reason that it proves an effective tool for the analysis and communication of ideas and arguments.

–A.A., San Francisco, October 2013

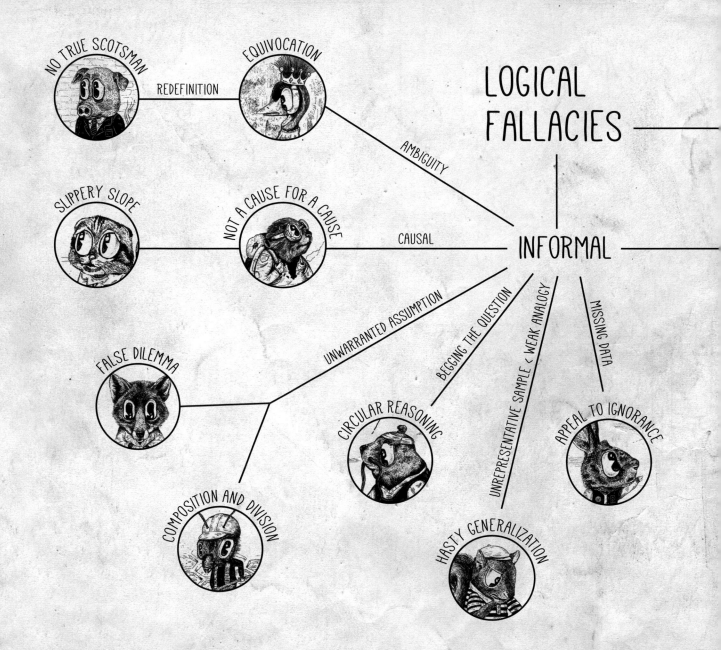

LOGICAL FALLACIES

INFORMAL

NO TRUE SCOTSMAN — REDEFINITION — EQUIVOCATION

EQUIVOCATION — AMBIGUITY — INFORMAL

SLIPPERY SLOPE — NOT A CAUSE FOR A CAUSE

NOT A CAUSE FOR A CAUSE — CAUSAL — INFORMAL

FALSE DILEMMA — UNWARRANTED ASSUMPTION

COMPOSITION AND DIVISION

CIRCULAR REASONING — BEGGING THE QUESTION

HASTY GENERALIZATION — UNREPRESENTATIVE SAMPLE < WEAK ANALOGY

APPEAL TO IGNORANCE — MISSING DATA

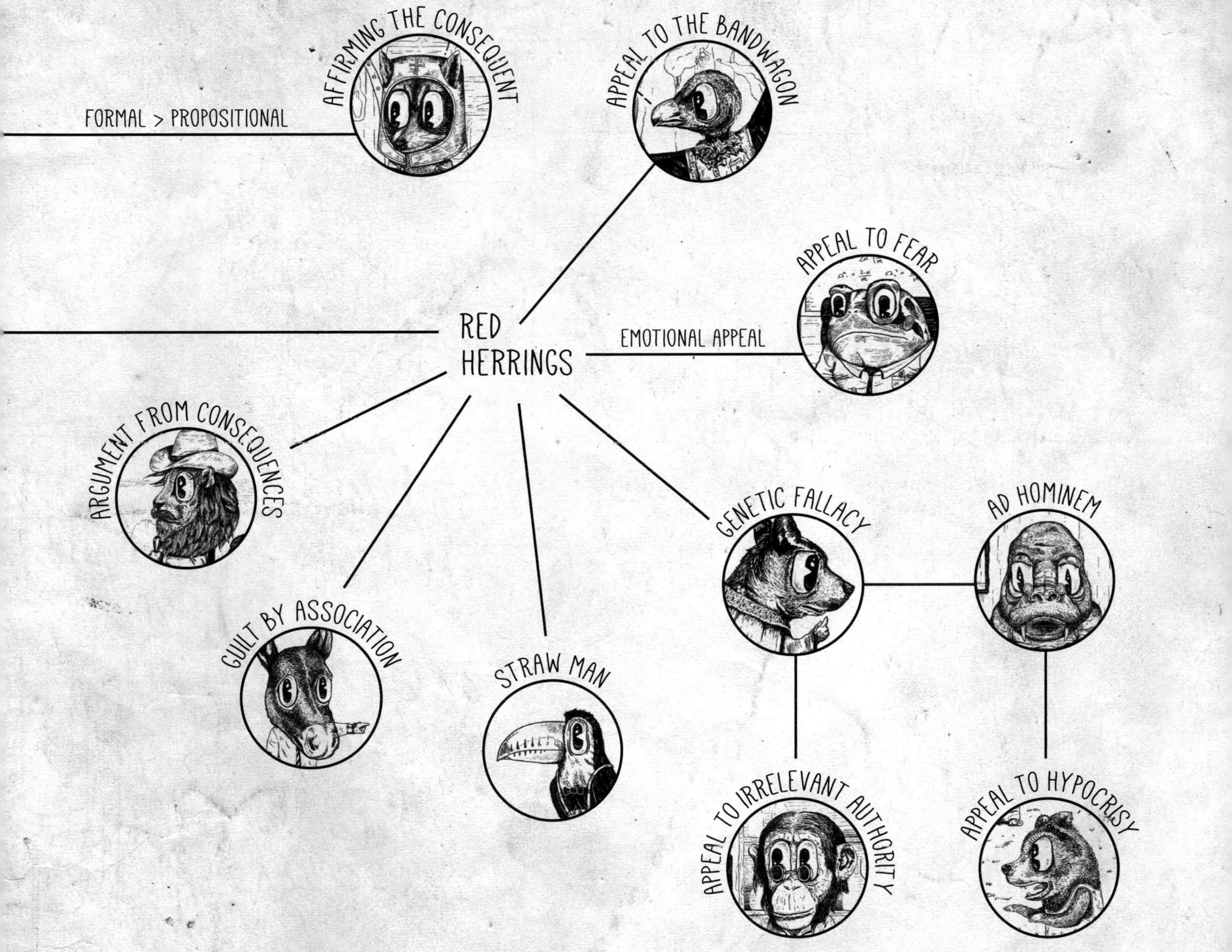

FORMAL > PROPOSITIONAL

AFFIRMING THE CONSEQUENT

APPEAL TO THE BANDWAGON

APPEAL TO FEAR

RED HERRINGS

EMOTIONAL APPEAL

ARGUMENT FROM CONSEQUENCES

GENETIC FALLACY

AD HOMINEM

GUILT BY ASSOCIATION

STRAW MAN

APPEAL TO IRRELEVANT AUTHORITY

APPEAL TO HYPOCRISY

ARGUMENT FROM CONSEQUENCES

Arguing from consequences is speaking for or against the truth of a statement by appealing to the consequences it would have *if* true (or if false). But the fact that a proposition leads to some unfavorable result does not mean that it is false. Similarly, just because a proposition has good consequences does not all of a sudden make it true. As history professor and author David Hackett Fischer puts it, "It does not follow that a quality which attaches to an effect is transferable to the cause" [Fischer].

In the case of good consequences, such an argument may appeal to an audience's hopes, which at times take the form of wishful thinking. In the case of bad consequences, the argument may instead play on an audience's fears. For example, take Dostoevsky's line, "If God does not exist, then everything is permitted." Discussions of objective morality aside, the apparent grim consequences of a purely materialistic world say nothing about whether or not it is true that God exists.

One should keep in mind that such arguments are faulty only when they are used to support or deny the *truth* of a statement, and not when they deal with decisions or policies [Curtis]. For example, a politician may logically oppose raising taxes for fear that it would adversely impact the lives of his constituents.

This fallacy is one of many in this book that can be termed a *red herring*, because it subtly redirects the discussion away from the original proposition—in this case, to the proposition's result.

WELL, IF WE GET RID OF OUR COWS, THEN WE WILL HAVE TO WALK EVERYWHERE, AND THAT WOULD BE TERRIBLE FOR MORALE. THEREFORE COW EMISSIONS ARE NOT KILLING OUR PLANET.

STRAW MAN

To "put up a straw man" is to intentionally caricature a person's argument with the aim of attacking the caricature rather than the actual argument. Misrepresenting, misquoting, misconstruing, and oversimplifying an opponent's position are all means by which one can commit this fallacy. The straw man argument is usually more absurd than the actual argument, making it an easier target to attack. It may also lure the other person toward defending the more ridiculous argument rather than their original one.

For example, a skeptic of Darwinism might say, "My opponent is trying to convince you that we evolved from chimpanzees who were swinging from trees, a truly ludicrous claim." This is a misrepresentation of what evolutionary biology actually claims, which is that humans and chimpanzees shared a common ancestor millions of years ago. Misrepresenting the idea is much easier than refuting the evidence for it.

THE ENERGETIC, MUSCULAR, AND COLORFUL TOUCAN WAS COMPLETELY
MISREPRESENTED BY ONE OF THE ARTISTS. LATER ON,
HE SHOWED THE AUDIENCE HIS PAINTING AND CRITICIZED
HOW DULL AND LIFELESS THE TOUCAN HAD LOOKED.

APPEAL TO IRRELEVANT AUTHORITY

An *appeal to authority* is an appeal to one's sense of modesty, which is to say, an appeal to the feeling that others are more knowledgeable [Engel], which may often—but of course not always—be true. One may reasonably appeal to *pertinent* authority, as scientists and academics typically do. A vast majority of the things that we believe in, such as atoms and the solar system, are on reliable authority, as are all historical statements, to paraphrase C. S. Lewis. An argument is more likely to be fallacious when the appeal is made to an *irrelevant* authority, one who is not an expert on the issue at hand. A similar appeal worth noting is the *appeal to vague authority*, where an idea is attributed to a faceless collective. For example, "Professors in Germany showed such and such to be true."

One type of appeal to irrelevant authority is the *appeal to ancient wisdom*, in which a belief is assumed to be true just because it originated some time ago. For example, "Astrology was practiced in ancient China, one of the most technologically advanced civilizations of the day." This type of appeal often overlooks the fact that some things are idiosyncratic and change naturally over time. For example, "We do not get enough sleep nowadays. Just a few centuries ago, people used to sleep for nine hours a night." There are all sorts of reasons why people might have slept longer in the past. The fact that they did is insufficient evidence for the argument that we should do so today.

PECULIARLY, PROFESSOR CHIMP, THE WORLD'S MOST DISTINGUISHED LIVING CHEMIST, IS OFTEN QUOTED ABOUT MATTERS OF FIDELITY.

EQUIVOCATION

Equivocation exploits the ambiguity of language by changing the meaning of a word during the course of an argument and using the different meanings to support an ill-founded conclusion.[2] (A word whose meaning is maintained throughout an argument is described as being used *uni*vocally.) Consider the following argument: "How can you be against faith when you take leaps of faith all the time: making investments, trusting friends, and even getting engaged?" Here, the meaning of the word "faith" is shifted from a spiritual belief in a creator to a willingness to undertake risks.

This fallacy is commonly invoked in discussions of science and religion, where the word "why" may be used equivocally. In one context, it is a word that seeks *cause*, which as it happens is the main driver of science, and in another it is a word that seeks *purpose*, which deals with morality and other realms where science may well have no answers. For example, one might argue: "Science cannot tell us why things are. Why do we exist? Why be moral? Thus, we need some other source to tell us why things happen."

[2] The illustration is based on an exchange between Alice and the White Queen in Lewis Carroll's *Through the Looking-Glass*.

THE QUEEN TOLD THE CURIOUS LITTLE CRANE THAT SHE COULD HAVE JAM EVERY OTHER DAY BUT NEVER TODAY, SINCE TODAY WAS NOT ANY OTHER DAY.

FALSE DILEMMA

A *false dilemma* is an argument that presents a limited set of two possible categories and assumes that everything in the scope of the discussion must be an element of that set.[3] Thus, by rejecting one category, you are forced to accept the other. For example, "In the war on fanaticism, there are no sidelines; you are either with us or with the fanatics." In reality, there is a third option, one could very well be neutral; and a fourth option, one may be against both; and even a fifth option, one may empathize with elements of both.

In *The Strangest Man*, Paul Dirac's biographer recounts a parable that physicist Ernest Rutherford once told his colleague Niels Bohr: A man bought a parrot from a pet store, only to bring it back because it didn't talk. After several such visits, the store manager eventually said, "Oh, that's right! You wanted a parrot that talks. Please forgive me. I gave you the parrot that thinks" [Farmelo]. Rutherford was clearly using the parable to illustrate the genius of the silent Dirac, but one can imagine how someone might use such a line of reasoning to suggest that a person is *either* silent and a thinker *or* talkative and an imbecile.

---

[3] This fallacy may also be referred to as the *fallacy of the excluded middle*, the *black and white fallacy* or a *false dichotomy*.

"WHICH PART OF THE AVOCADO WOULD YOU LIKE TO TRY?"
SAID THE MERCHANT.
"I WANT TO TRY THE MIDDLE BIT" SAID THE BUYER,
"WHICH SEEMS TO BE MISSING".

NOT A CAUSE FOR A CAUSE

This fallacy assumes a cause for an event where there is no evidence that one exists. When two events occur one after the other (or simultaneously), this may be by coincidence, or due to some other unknown factor. One cannot conclude that one event *caused* the other without evidence. "The recent earthquake was because we disobeyed the king" is not a good argument.

This fallacy has two specific types: "after this, therefore because of this" (*post hoc ergo propter hoc*) and "with this, therefore because of this" (*cum hoc ergo propter hoc*). With the former, because one event preceded another, it is said to have been the cause. With the latter, because an event happened at the same time as another, it is said to have been the cause. In various disciplines, this is known as confusing *correlation* with *causation*.[4]

Here is an example paraphrased from comedian Stewart Lee: "I can't say that, because in 1976 I did a drawing of a robot and then *Star Wars* came out, they must have copied the idea from me." And here is another that I recently saw on an online forum: "The hacker took down the railway company's website, and when I checked the train schedule, what do you know, they were all delayed!" What the poster failed to realize is that trains can be late for all kinds of reasons, so without any kind of scientific control, the inference that the hacker was the cause is unfounded.

---

[4] As it turns out, eating chocolate and winning a Nobel Prize have been shown to be highly correlated, perhaps raising the hopes of many a chocolate eater [Pritchard]: bbc.co.uk/news/magazine-20356613

AT THE END OF EVERY NIGHT AND SHORTLY BEFORE DAWN, THE BEAVER WALKS ALL THE WAY TO THE TOP OF THE MOUNTAIN AND ASKS THE SUN TO COME OUT. THE SUN ALWAYS DOES.

APPEAL TO FEAR

This fallacy plays on the fears of an audience by imagining a scary future that would be of their making if some proposition were accepted. Rather than provide solid evidence that the proposition would lead to a certain conclusion (which might be a legitimate cause for fear), such arguments rely on rhetoric, threats, or outright lies. For example, "I ask all employees to vote for my chosen candidate in the upcoming election. If the other candidate wins, he will raise taxes and many of you will lose your jobs."

Here is another example, drawn from the novel *The Trial*: "You should give me all your valuables before the police get here. They will end up putting them in the storeroom, and things tend to get lost in the storeroom." Here, although the argument is more likely a threat, albeit a subtle one, an attempt is made at reasoning. Blatant threats or orders that do *not* attempt to provide evidence should not be confused with this fallacy, even if they exploit one's sense of fear [Engel].

When an appeal to fear proceeds to describe a series of terrifying events that will occur as a result of accepting a proposition—without clear causal links between them—it becomes reminiscent of a *slippery slope* argument. And when the person making the appeal provides one and only one alternative to the proposition under attack, it becomes reminiscent of a *false dilemma*.

MR. FROG LOST THE ELECTION AFTER MR. DONKEY CONVINCED EVERYONE THAT IF MR. FROG BECAME THE SCHOOL DEAN, SOON ENOUGH, THE ENTIRE UNIVERSITY WOULD BE RUN BY FROGS.

HASTY GENERALIZATION

This fallacy is committed when one forms a conclusion from a sample that is either too small or too special to be representative. For example, asking ten people on the street what they think of the president's plan to reduce the deficit can in no way be said to gauge the sentiment of the entire nation.

Although convenient, hasty generalizations can lead to costly and catastrophic results. For instance, it may be argued that an engineering assumption led to the explosion of the *Ariane 5* rocket during its first test flight: The control software had been extensively tested with the previous model, *Ariane 4*—but unfortunately these tests did not cover all the possible scenarios of the *Ariane 5*, so it was wrong to assume that the data would carry over. Signing off on such decisions typically comes down to engineers' and managers' ability to argue, hence the relevance of this and similar examples to our discussion of logical fallacies.

There is another example in *Alice's Adventures in Wonderland*, where Alice infers that, since she is floating in a body of water, a railway station, and thus help, must be close by: "Alice had been to the seaside once in her life, and had come to the general conclusion, that wherever you go to on the English coast you find a number of bathing machines in the sea, some children digging in the sand with wooden spades, then a row of lodging houses, and behind them a railway station" [Carroll].

"I HAVE NEVER SEEN FOOD THAT IS NOT CIRCLE-SHAPED. ALL FOOD IS THEREFORE DEFINITELY CIRCLE-SHAPED."

"I HAVE NEVER SEEN FOOD THAT DOES NOT HAVE STRAIGHT EDGES..."

APPEAL TO IGNORANCE

This kind of argument assumes a proposition to be true simply because there is no evidence proving that it is *false*.[5] Hence, absence of evidence is taken to be evidence of absence. Carl Sagan gives this example: "There is no compelling evidence that UFOs are not visiting the Earth; therefore UFOs exist" [Sagan]. Similarly, before we knew how the pyramids were built, some concluded that, unless proven otherwise, they must have been built by a supernatural power. But in fact, the "burden of proof" always lies with the person making a claim.

More logically, and as several others have put it, one should ask what is likely based on evidence from past observation. Which is more likely: That an object flying through space is a man-made artifact or natural phenomenon, or that it is aliens visiting from another planet? Since we have frequently observed the former and never the latter, it is more reasonable to conclude that UFOs are probably *not* aliens visiting from outer space.

A specific form of the appeal to ignorance is the *argument from personal incredulity*, where a person's inability to imagine something leads them to believe that it is false. For example, "It is impossible to imagine that we actually landed a man on the moon, therefore it never happened." Responses of this sort are sometimes wittily countered with, "That's why you're not a physicist!"

---

[5] The illustration is inspired by Neil deGrasse Tyson's response to an audience member's question on UFOs: bookofbadarguments.com/video/tyson

LOOK! IT'S A STRANGE BEAM OF LIGHT MOVING THROUGH THE SKY. I DON'T KNOW WHAT IT IS, SO IT MUST BE ALIENS VISITING US FROM ANOTHER PLANEt.

NO TRUE SCOTSMAN

This argument comes up after someone has made a general claim about a group of things, and then been presented with evidence challenging that claim. Rather than revising their position, or contesting the evidence, they dodge the challenge by arbitrarily redefining the criteria for membership in that group.[6]

For example, someone may posit that programmers are creatures with no social skills. If someone else comes along and repudiates that claim by saying, "But John is a programmer, and he is not socially awkward at all," this may provoke the response, "Yes, but John isn't a *true* programmer." Here, it is not clear what the attributes of a programmer are; the category is not as clearly defined as that of, say, people with blue eyes. The ambiguity allows the stubborn mind to redefine things at will.

This fallacy was coined by Antony Flew in his book *Thinking about Thinking*. There, he gives the following example: Hamish is reading the newspaper and comes across a story about an Englishman who has committed a heinous crime, to which he reacts by saying, "No Scotsman would do such a thing." The next day, he comes across a story about a Scotsman who has committed an even worse crime. Instead of amending his claim about Scotsmen, he reacts by saying, "No *true* Scotsman would do such a thing" [Flew].

[6] When an attacker maliciously redefines a category, knowing well that by doing so, he or she is intentionally misrepresenting it, the attack becomes reminiscent of the *straw man* fallacy.

GENETIC FALLACY

A *genetic fallacy* is committed when an argument is either devalued or defended solely because of its origins. In fact, an argument's history or the origins of the person making it have no effect whatsoever on its validity. As T. Edward Damer points out, when one is emotionally attached to an idea's origins, it is not always easy to disregard those feelings when evaluating the argument's merit [Damer].

Consider the following argument: "Of course he supports the union workers on strike; he is, after all, from the same village." Here, the argument supporting the workers is not being evaluated based on its merits; rather, because the person behind it happens to come from the same village as the protesters, we are led to infer that his position is worthless. Here is another example: "As men and women living in the twenty-first century, we cannot continue to hold these Bronze Age beliefs." Why not, one might ask. Are we to dismiss all ideas that originated in the Bronze Age simply because they came about at that time?

Conversely, one may also invoke the genetic fallacy in a positive sense, by saying, for example, "Jack's views on art cannot be contested; he comes from a long line of eminent artists." Here, the evidence used for the inference is as lacking as in the previous examples.

GUILT BY ASSOCIATION

Guilt by association is used to discredit an argument for proposing an idea that is shared by some socially demonized individual or group. For example, "My opponent is calling for a healthcare system that would resemble that of socialist countries. Clearly, that would be unacceptable." Whether or not the proposed healthcare system resembles that of socialist countries has no bearing whatsoever on whether it is good or bad; it is a complete non sequitur.

   Another argument, which has been repeated ad nauseam in some societies, is this: "We cannot let women drive cars because people in godless countries let their women drive cars." Essentially, what these examples try to argue is that some group of people is absolutely and categorically bad. Hence, sharing even a single attribute with that group would make one a member of it, which would then bestow on one all the evils associated with that group.

A PROSPEROUS DICTATORSHIP
IS AN EDUCATED DICTATORSHIP

MY OPPONENT BELIEVES THAT WE SHOULD SPEND MORE ON EDUCATION.
DO YOU KNOW WHO ELSE THINKS THAT? THE DICTATOR HIMSELF!!!

AFFIRMING THE CONSEQUENT

One of several valid formal arguments is known as *modus ponens* (the mode of affirming) and takes the following form: **If A then C, A; hence C.** More formally: $A \Rightarrow C, A \vdash C$. A is called the *antecedent* and C the *consequent*, and they form two premisses and a conclusion. For example:

Premiss: If A then C

If water is boiling at sea level, then its temperature is at least 100°C.
This water is boiling at sea level; hence its temperature is at least 100°C.

Premiss: A   **Conclusion: C**

Such an argument is sound in addition to being valid.

Affirming the consequent is a formal fallacy that takes this form: **If A then C, C; hence A.** The error lies in assuming that because the consequent is true, the antecedent must also be true, which in reality need not be the case.

For example, "People who go to college are successful. John is successful, hence he must have gone to college." Clearly, John's success *could* be a result of schooling, but it could also be a result of his upbringing, or perhaps his eagerness to overcome difficult circumstances. Generally, because schooling is not the *only* path to success, one cannot say that a person who is successful *must* have received schooling.

APPEAL TO HYPOCRISY

Also known by its Latin name, *tu quoque*, meaning "you too," this fallacy involves countering someone's argument by pointing out that it conflicts with his or her own past actions or statements [Engel]. Thus, by answering a charge with a charge, it diverts attention from the argument at hand to the person making it. This characteristic makes the fallacy a particular type of *ad hominem* attack. Of course, just because someone has been inconsistent about his position does not mean that his position cannot be correct.

On an episode of the topical British TV show *Have I Got News for You*, a panelist objected to a protest in London against corporate greed because of the protesters' apparent hypocrisy, pointing out that while they professed to be against capitalism, they continued to use smartphones and buy coffee.[7]

Here is another example, from Jason Reitman's movie *Thank You for Smoking*, where a *tu quoque*–laden exchange is ended by the smooth-talking tobacco lobbyist Nick Naylor: "I'm just tickled by the idea of the gentleman from Vermont calling me a hypocrite when this same man, in one day, held a press conference where he called for the American tobacco fields to be slashed and burned, then he jumped on a private jet and flew down to Farm Aid where he rode a tractor onstage as he bemoaned the downfall of the American farmer."

[7] That excerpt is available here: bookofbadarguments.com/video/hignfy

SLIPPERY SLOPE

A *slippery slope* argument attempts to discredit a proposition by arguing that its acceptance will undoubtedly lead to a sequence of events, one or more of which are undesirable.[8] Although the sequence of events may be *possible*—each transition occurring with some probability—this type of argument assumes that every transition is *inevitable*—while providing no evidence in support of that. This fallacy plays on the fears of an audience and is related to a number of other fallacies, such as the *appeal to fear*, the *false dilemma*, and the *argument from consequences*.

   For example, "We shouldn't allow people uncontrolled access to the internet. The next thing you know they will be frequenting pornographic websites, and soon enough, our entire moral fabric will disintegrate and we will be reduced to animals." As is glaringly clear, no evidence is given, other than unfounded conjecture, that internet access implies the disintegration of a society's moral fabric. Moreover, the argument presupposes certain things about people's behavior within the society.

---

[8] The slippery slope fallacy described here is of a causal type.

Informal Fallacy › Causal Fallacy › Not a Cause for a Cause › **Slippery Slope**

APPEAL TO THE BANDWAGON

Also known as the *appeal to the people*, this argument uses the fact that many people (or even a majority) *believe* in something as evidence that it must be true. This type of argument has often impeded the widespread acceptance of a pioneering idea. For example, most people in Galileo's day believed that the sun and the planets orbited around Earth, so Galileo faced ridicule for his support of the Copernican model, which correctly puts the sun at the center of our solar system. More recently, physician Barry Marshall had to take the extreme measure of dosing himself with *H. pylori* bacteria in order to convince the scientific community that it may cause peptic ulcers, a theory that was, initially, widely dismissed.

Advertisements frequently use this method to lure people into accepting something solely because it is popular. For example, "All the cool kids use this hair gel; be one of them." Although becoming a "cool kid" is an enticing offer, it does nothing to support the imperative that one should buy the advertised product. Politicians also use similar rhetoric to add momentum to their campaigns and influence voters.

AD HOMINEM

An *ad hominem* argument (from the Latin for "to the man") is one that attacks a person rather than the argument he or she is making, with the intention of diverting the discussion and discrediting their argument.[9] For example, "You're not a historian; why don't you stick to your own field?" Here, the fact that someone is not a historian has no impact on the merit of their argument (since, of course, it is not the case that anyone other than a historian is automatically wrong on the subject), so it does nothing to strengthen the attacker's position.

This type of personal attack is referred to as *abusive* ad hominem. A second type, *circumstantial* ad hominem, attacks a person for cynical reasons, usually by making a judgment about their intentions. For example, "You don't really care about lowering crime in the city; you just want people to vote for you." But *even if* a person would benefit from their argument's acceptance, this does not mean they must be wrong.

An *ad hominem* attack sometimes succeeds at changing the subject by devolving into a *tu quoque* exchange. For example, John says, "This man is wrong because he has no integrity; just ask him why he was fired from his last job," to which Jack replies, "How about we talk about the fat bonus you took home last year despite half your company being downsized," by which point the discussion has gone completely off track. That said, there are situations where one may legitimately question a person's credibility, such as during trial testimony.

[9] The illustration is inspired by a discussion on Usenet several years ago in which an overzealous and stubborn programmer was a participant.

Informal Fallacy › Red Herring › Genetic Fallacy › **Ad Hominem**

"YOUR AD HOMINEM ATTACKS ARE EVIDENCE THAT
YOUR ARGUMENTS ARE BASELESS" WROTE USER 226.
RODNEY BEGAN TYPING HIS REPLY: "YOU APPEAR TO BE
TOO STUPID TO UNDERSTAND THE DIFFERENCE BETWEEN AN
INSULT AND AN AD HOMINEM ATTACK"

CIRCULAR REASONING

Circular reasoning is one of four types of arguments known as *begging the question*, [Damer] where one implicitly or explicitly assumes the conclusion in one or more of the premisses. In circular reasoning, a conclusion is either blatantly used as a premiss, or more often, it is reworded to appear as though it is a different proposition when in fact it is not. For example, "You're utterly wrong because you're not making any sense." Here, the two propositions are one and the same, since being wrong and not making any sense mean the same thing in this context. The argument is simply stating "Because of x therefore x," which is meaningless.

A circular argument may at times rely on unstated premisses, which can make it more difficult to detect. Consider someone who tells an atheist that he should believe in God because otherwise he will go to hell. The unstated premiss behind anyone going to hell is that there exists a God to send him there. Hence, the premiss "There exists a God who sends nonbelievers to hell" is used to support the conclusion "There exists a God." As comedian Josh Thomas tells Peg on the Australian TV series *Please Like Me*, "You can't threaten an atheist with hell, Peg. It doesn't make any sense. It's like a hippie threatening to punch you in your aura."

COMPOSITION AND DIVISION

One commits the *fallacy of composition* by inferring that, because the parts of a whole have a particular attribute, the whole must have that attribute also. But to paraphrase Peter Millican, if every sheep in a flock has a mother, it does not then follow that the *flock* has a mother. Here is another example: "Each module in this software system has been subjected to a set of unit tests and passed them all. Therefore, when the modules are integrated, the software system will not violate any of the invariants verified by those unit tests." The reality is that putting individual parts together to form a system introduces a new level of complexity, due to how the parts interact, which may in turn introduce new ways for things to go wrong.

Conversely, to commit the *fallacy of division* is to infer that part of a whole must have some attribute because the whole to which it belongs happens to have that attribute. For example, "Our team is unbeatable. Any one of our players would be able to take on a player from the other team and outshine him." While it may be true that the team as a whole is unbeatable, this could well be the result of how the players' individual skills work *together*—so one cannot use this as evidence that each player is unbeatable on their own.

FINAL REMARKS

Many years ago, I heard a professor introduce deductive arguments using a wonderful metaphor, describing them as watertight pipes where truth goes in one end and truth comes out the other end. As it happens, that was the inspiration for this book's cover. Having reached the end of this book, I hope that you leave with not only a better appreciation of the benefits of watertight arguments in validating and expanding knowledge, but also of the complexities of inductive arguments where probability comes into play. With the latter in particular, critical thinking proves an indispensable tool. Most importantly, I hope that you leave with a greater awareness of the dangers of flimsy arguments and how commonplace they are in our everyday lives.

• • •

I find it only appropriate to conclude by thanking the people with whom I have had the pleasure to watch this project go from its embryonic stage to flight mode: Thanks to everyone who took the time to send in their comments and critiques (feedback that undoubtedly improved this book); the 700,000 readers of the online edition; the nearly 4,000 readers who supported the project with donations or by purchasing the first edition; the bookshops that were kind enough to stock the first edition, obscure as it was; and especially the volunteers who have translated the online edition into their own languages. It has been a wonderful journey, and I trust that it is but one of many to come.

ARGUMENT: A set of *propositions* aimed at persuading through reasoning. In an argument, a subset of propositions, called *premisses*, provides support for some other proposition called the conclusion.

proposition: A statement that is either true or false, but not both. For example, "Boston is the largest city in Massachusetts."

premiss: A *proposition* that provides support to an *argument*'s conclusion. An argument may have one or more premisses. Also spelled "premise."

falsifiable: A *proposition* or *argument* is falsifiable if it can be refuted, or disproved, through observation or experiment. For example, the proposition "All leaves are green" may be refuted by pointing to a leaf that is not green. Falsifiability is a sign of an argument's strength, rather than of its weakness.

LOGICAL FALLACY: An error in the reasoning used to transition from one *proposition* to the next, which results in a faulty *argument*. Logical fallacies violate one or more of the principles that make a good argument, such as good structure, consistency, clarity, order, relevance, and completeness. It is important to note that finding a fallacy in an argument is not the same as proving its conclusion false—the conclusion may be true, but need better reasoning to back it up.

formal fallacy: An error in reasoning that is illogical because its structure is faulty. The fallacy can be spotted just by analyzing the argument's form, without needing to evaluate its content. (For example, see *affirming the consequent* on page 32.)

informal fallacy: An error in reasoning that is illogical due to its content and context rather than its form. The error ought to be a commonly invoked one to be considered an informal fallacy. (Nearly all of the fallacies in this book are informal.)

DEDUCTIVE ARGUMENT: An argument in which if the premisses are true, then the conclusion must be true. The conclusion is said to follow with logical necessity from the premisses. For example, "All men are mortal. Socrates is a man. Therefore, Socrates is mortal." A deductive argument is intended to be *valid*, but of course might not be.

valid: A *deductive argument* is valid if its conclusion does in fact follow logically from its *premisses*. Otherwise, it is said to be invalid. The descriptors "valid" and "invalid" apply only to *arguments* and not to *propositions*.

sound: A *deductive argument* is sound if it is *valid* and its *premisses* are true. If either of those conditions does not hold, then the argument is unsound. Truth is determined by looking at whether the argument's premisses and conclusions are in accordance with facts in the real world.

INDUCTIVE ARGUMENT: An *argument* in which if the *premisses* are true, then it is probable that the conclusion will also be true.[10] The conclusion does not follow from the premisses with logical necessity, but rather with probability. For example, "Every time we measure the speed of light in a vacuum, it is 3×10^8 m/s. Therefore, the speed of light in a vacuum is a universal constant." Inductive arguments usually proceed from specific instances to the general.

strong: An *inductive argument* is strong if, in the case that its *premisses* are true, then it is highly probable that its conclusion is also true. Otherwise, if it is improbable that its conclusion is true, then it is said to be weak. Because they rely on probability, inductive arguments are not truth-preserving; it is never the case that a true conclusion must follow from true premisses.

cogent: An *inductive argument* is cogent if it is *strong* and the *premisses* are actually true—that is, in accordance with facts. Otherwise, it is said to be uncogent.

---

[10] In science, one usually proceeds inductively from data to laws to theories, hence induction is the foundation of much of science. Induction is typically taken to mean either testing a proposition on a sample (because it would be impractical to test more extensively) or using reason alone (because it is impossible to test at all).

BIBLIOGRAPHY

Aristotle. *On Sophistical Refutations*. Trans. W. A. Pickard-Cambridge.
http://classics.mit.edu/Aristotle/sophist_refut.html.

Avicenna. *Avicenna's Treatise on Logic*. Trans. and ed. Farhang Zabeeh.
The Hauge: Nijhoff, 1971.

Carroll, Lewis. *Alice's Adventures in Wonderland*.
www.gutenberg.org/files/11/11-h/11-h.htm.

Curtis, Gary N. Fallacy Files. http://fallacyfiles.org.

Damer, T. Edward. *Attacking Faulty Reasoning: A Practical Guide to Fallacy-Free
Arguments*. 6th ed. Belmont, CA: Wadsworth Cengage Learning, 2009.

Engel, S. Morris. *With Good Reason: An Introduction to Informal Fallacies*.
Boston: Bedford/St. Martin's, 1999.

Farmelo, Graham. *The Strangest Man: The Hidden Life of Paul Dirac,
Mystic of the Atom*. New York: Basic Books, 2011.

Fieser, James. Internet Encyclopedia of Philosophy. www.iep.utm.edu.

Firestein, Stuart. *Ignorance: How It Drives Science*.
Oxford: Oxford Univ. Press, 2012.

Fischer, David Hackett. *Historians' Fallacies: Toward a Logic of Historical Thought.* New York: Harper & Row, 1970.

Flew, Antony. *Thinking about Thinking.* Glasgow: Fontana/Collins, 1975.

Gula, Robert J. *Nonsense: A Handbook of Logical Fallacies.* Mount Jackson, VA: Axios Press, 2002.

Hamblin, Charles. *Fallacies.* London: Methuen, 1970.

King, Stephen. *On Writing: A Memoir of the Craft.* New York: Scribner, 2000.

Minsky, Marvin. *The Society of Mind.* New York: Simon & Schuster, 1988.

Pólya, George. *How to Solve It: A New Aspect of Mathematical Method.* Princeton: Princeton Univ. Press, 2004.

Pritchard, Charlotte. "Does Chocolate Make You Clever?" *BBC News Magazine.* November 19, 2012. http://bbc.co.uk/news/magazine-20356613.

Russell, Bertrand. *The Problems of Philosophy.* London: Williams & Norgate, 1912. http://ditext.com/russell/russell.html.

Sagan, Carl. *The Demon-Haunted World: Science as a Candle in the Dark.* New York: Random House, 1995.

Simanek, Donald E. Uses and Misuses of Logic. http://lhup.edu/~dsimanek/philosop/logic.htm.

Smith, Peter. *An Introduction to Formal Logic.* Cambridge: Cambridge Univ. Press, 2003.

ABOUT THE AUTHOR AND ILLUSTRATOR

Ali Almossawi holds a Masters in Engineering Systems from the Massachusetts Institute of Technology (MIT) and a Masters in Software Engineering from Carnegie Mellon University. He resides in San Francisco with his wife and daughter, where he works as a data visualization designer on Mozilla's Metrics Team, and occasionally collaborates with the MIT Media Lab. Formerly, Ali spent time at both Harvard and the Software Engineering Institute (SEI), where his research involved creating predictive models of source code quality. His work has appeared in *Scientific American*, *Wired*, *The New York Times*, *Fast Company* and others.

Almossawi.com

Alejandro Giraldo holds a Graphic Design degree from UPB Medellín and an MA in Art Direction from ELISAVA (the Barcelona School of Design and Engineering). He lives in Medellín, Colombia, where he works on various freelance projects.

AlejoGiraldo.com

Cheer up this bear . . .
visit BookofBadArguments.com